The Geosphere

Illustrations: Janet Moneymaker
Design/Editing: Marjie Bassler

The Geosphere
ISBN 978-1-953542-17-5

Published by Gravitas Publications Inc.
Imprint: Real Science-4-Kids
www.gravitaspublications.com
www.realscience4kids.com

The **geosphere** is the part of Earth made of rocks, minerals, and soils.

Rocks!

The geosphere goes from the **crust** on the outside of Earth all the way to the **core** in the center of Earth.

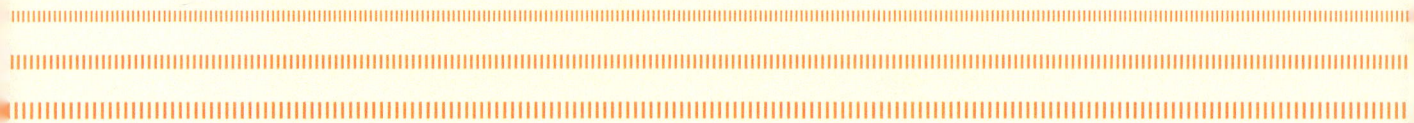

Earth is made of different layers.

The outer layer is the **crust.**

The middle layer is the **mantle.**

The innermost layer is the **core.**

The **crust** is solid.

The **mantle** is solid on top and soft like peanut butter below.

The **core** is soft metal on the outside and solid metal in the middle.

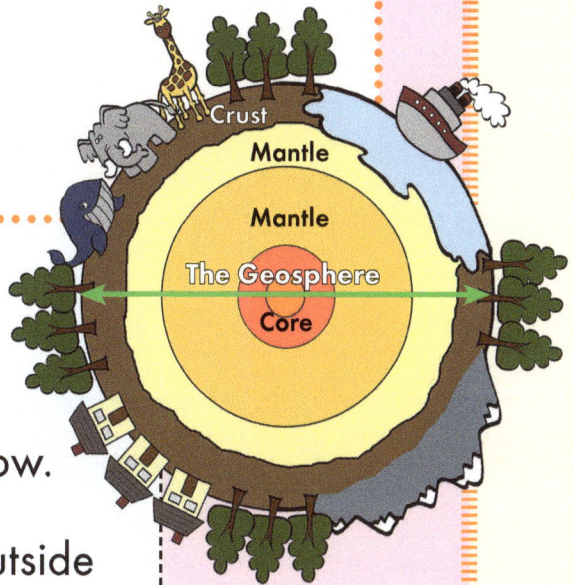

Crust

Mantle

Mantle

The Geosphere

Core

The geosphere is shaped and changed by both big and small events.

Volcanoes and **earthquakes** are big events.

Volcanoes change the Earth by creating mountains and other landforms from magma that has come to the surface and cooled.

Earthquakes change Earth when the land shifts and valleys, waterways, and mountains are formed.

Volcanoes occur when soft, hot magma inside the Earth is forced through a weak area in the mantle and crust.

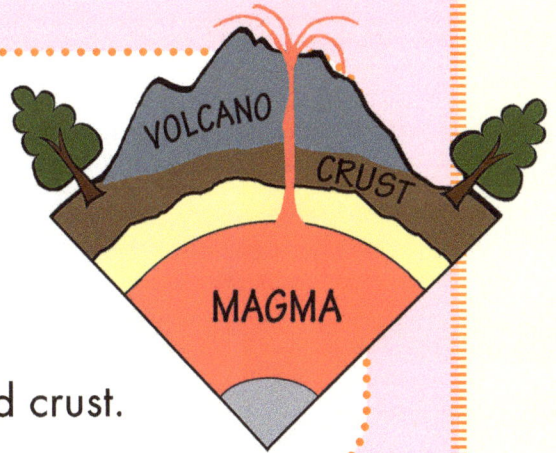

VOLCANO

CRUST

MAGMA

Earthquakes occur when parts of the crust and upper mantle suddenly shift. This movement can create mountains and valleys.

Wind changes the geosphere when it blows sand around to create shifting sand dunes.

Amazing!

Earth is also shaped and changed by **erosion**. Erosion of rocks and soil occurs when they are worn away, crumbled, or ground down by water, wind, or ice.

Look what water can do!

Powerful!

A **glacier** is a huge river of ice that flows very slowly down a slope. The glacier carries rocks along beneath it that grind down the land underneath the glacier.

Glaciers can make HUGE valleys!

Yes. And that takes a huge amount of time.

Animals change the geosphere when they dig in the Earth to make homes. Animals also make trails to walk on and sometimes block streams by building dams.

The geosphere gives us resources that we use to power cars and batteries and to heat our homes.

Oil, gas, and minerals.

Studying the geosphere helps us learn about the history of Earth. Layers of rocks in the crust show us how landscapes were formed. And **fossils** show us what life forms existed in the ancient past. Fossils are remains or imprints of plants and animals that have been preserved in ancient rock.

How to say science words

core   (KAWR)

crust   (KRUHST)

earthquake   (ERTH-kwayk)

erosion   (i-ROH-zhuhn)

fossil   (FAH-suhl)

geosphere   (JEE-oh-sfeer)

glacier   (GLAY-shuhr)

mantle   (MAN-tuhl)

mineral   (MIN-ruhl)

science   (SIY-ens)

volcano   (vahl-KAY-noh)

What questions do you have about
THE GEOSPHERE?

Learn More Real Science!

Complete science curricula from Real Science-4-Kids

Focus On Series

Unit study for elementary and middle school levels

Chemistry
Biology
Physics
Geology
Astronomy

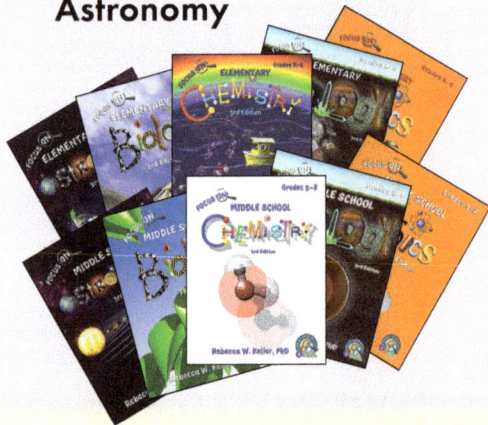

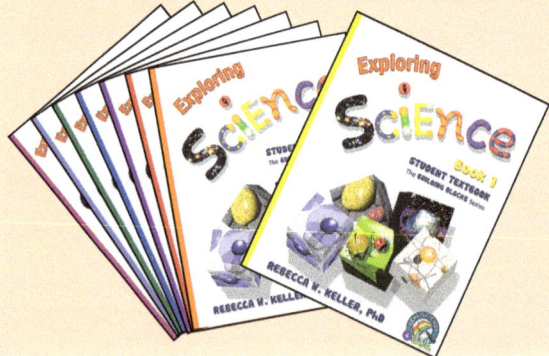

Exploring Science Series

Graded series for levels K–8. Each book contains 4 chapters of:

Chemistry
Biology
Physics
Geology
Astronomy